curious about

UPCYCLING

BY AMY S. HANSEN

AMICUS LEARNING

What are you

CHAPTER ONE
1

A New Life for Trash
PAGE **4**

CHAPTER TWO
2

A Closer Look
PAGE **10**

curious about?

CHAPTER THREE

What Can I Do?
PAGE **16**

Stay Curious! Learn More . . . 22
Glossary 24
Index 24

Curious About is published
by Amicus Learning,
an imprint of Amicus.
P.O. Box 227
Mankato, MN 56002
www.amicuspublishing.us

Copyright © 2025 Amicus. International copyright reserved in all countries. No part of this book may be reproduced in any form without written permission from the publisher.

Editor: Alissa Thielges
Series Designer: Kathleen Petelinsek
Book Designer and Photo Researcher: Emily Dietz

Library of Congress Cataloging-in-Publication Data
Names: Hansen, Amy, author.
Title: Curious about upcycling / by Amy S. Hansen.
Description: Mankato, MN : Amicus Learning, [2025] | Series: Curious about green living | Includes bibliographical references and index. | Audience: Ages 5–9 | Audience: Grades 2–3 | Summary: "What is upcycling and how can it help save the Earth? Ignite kids' growing curiosity about the environment with an inquiry-based approach to reducing waste. Includes infographics and back matter to support research skills, plus table of contents, glossary, and index"—Provided by publisher.
Identifiers: LCCN 2023043274 (print) | LCCN 2023043275 (ebook) | ISBN 9781645497004 (library binding) | ISBN 9781681529691 (paperback) | ISBN 9781645497066 (ebook)
Subjects: LCSH: Upcycling (Waste, etc.)—Juvenile literature. | Waste minimization—Juvenile literature.
Classification: LCC TD794.9 .H36 2025 (print) | LCC TD794.9 (ebook) | DDC 363.72/82—dc23/eng/20231205
LC record available at https://lccn.loc.gov/2023043274
LC ebook record available at https://lccn.loc.gov/2023043275

Photo Credits: Adobe Stock: almaholm 11 (glass bottle); Alamy: WENN Rights Ltd 12; Deposit Photos: vladymyr cover, 1; Dreamstime: Harry Wedzinga 11 (seeds), Valentin Ivantsov 5; Getty: AzmanL 2 (l), 8–9, Caiaimage/Sam Edwards 10, DarioEgidi 6, Eurngkwan 11 (vase), NickyLloyd 2 (r), 14–15, RecycleMan 11 (plastic bottle), South_agency 3, 20–21; Noun Project: 0733, 22 & 23 (brush), Imam ID 22 & 23 (bottle), Weno NL 7 (b); Shutterstock: Abdul Hakim Nurmaulana 11 (crayons), adriaticfoto 7 (t), Daisy Daisy 17, Kristine Rad 11 (feeder), Lambros Kazan 11 (art), Netrun78, 13, Sirle Kabanen 11 (tubes), Tatiana Buzmakova 18–19

Printed in China

CHAPTER ONE

What is upcycling?

Upcycling makes something useful out of trash. Take an old shirt and make a bag. Or create a game out of empty cereal boxes. That's upcycling. Most things follow a downward cycle. **Products** are created, bought, used, and then thrown away. Upcycling moves items back up in the cycle. That's how it got its name!

Upcycling gives new life to things that would have been thrown away.

A NEW LIFE FOR TRASH

A machine shreds paper to recycle.

Is upcycling like recycling?

It's similar! For both, you are reusing **materials**. But **recycling** takes many people and big machines. When we recycle, the metal or paper is broken down into its basic form. A new product is made. You can upcycle on your own. You find new ways to use the same product.

You can use empty cans and egg cartons as plant starters.

A NEW LIFE FOR TRASH

REDUCE
limit the number of materials we use

RECYCLE
break down materials to make new products

REUSE
give old items new purpose; avoid one-use products

A NEW LIFE FOR TRASH

Is upcycling better than recycling?

No. Both are important. We cannot upcycle every can, bottle, or piece of cardboard. But we can reuse some of them. And then we recycle the rest. We may even do both. Upcycle a cardboard box by turning it into a playhouse. Then recycle it later.

A family makes a playhouse out of a large cardboard box.

CHAPTER TWO

Where can I find materials to upcycle?

A CLOSER LOOK

You can check out garage sales for items to upcycle.

Look around your home! And get creative. Make a jump rope out of **plastic** bags. Glass jars can hold pencils or snacks. A peanut butter jar can become a snow globe. Check out **thrift stores** and garage sales for items, too. Or you can ask your friends.

WHAT TO UPCYCLE?

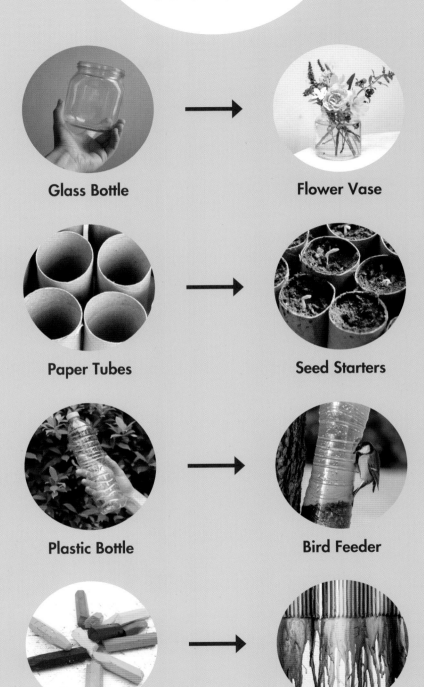

A CLOSER LOOK

A CLOSER LOOK

Why doesn't everyone upcycle?

Not everyone has the time. It's easier to buy something new. Sometimes, people start a project but don't finish it. Or they don't like how it turned out. Then they start over with new materials. Or they give up and throw it away.

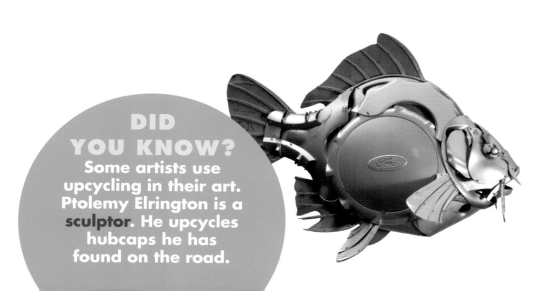

DID YOU KNOW?
Some artists use upcycling in their art. Ptolemy Elrington is a sculptor. He upcycles hubcaps he has found on the road.

This project uses an old sweater to make Christmas ornaments.

A CLOSER LOOK

A worker updates an old chair to sell.

A CLOSER LOOK

Do people make money upcycling?

DID YOU KNOW?
One boy in Virginia makes dog toys out of old t-shirts and tennis balls. His business has won awards.

A CLOSER LOOK

Some people do. They are good at imagining new uses for materials. Then they sell their creations. Some of those people **reclaim** furniture. Other people make jewelry or clothes. Often, they sell their items at festivals or online stores.

CHAPTER THREE

What skills do I need to start upcycling?

Each project needs different skills. Some people learn to paint. Others learn how to sew. For example, you can turn a water bottle into a planter. To do that, you need to cut plastic. Then paint the bottle. Then you need to take care of plants.

A boy hangs his water bottle planter in a tree.

What if I like buying new things?

A girl uses scrap fabric to make new doll clothes.

WHAT CAN I DO?

You can still buy new things. But you may find a new use for your old things. If a pair of pants becomes too short, cut them into shorts. Or make a quilt from old clothes. Upcycling keeps these clothes out of **landfills**. And you may like making things.

A boy creates a robot with empty food containers.

WHAT CAN I DO?

What can I do today?

WHAT CAN I DO?

Find something that is ready to be thrown away. Save an old shoe box from the garbage. You can build a home for your stuffed animals. Start an upcycling club with your friends. Upcycling lets you be creative!

STAY CURIOUS!

ASK MORE QUESTIONS

Where can I get ideas for upcycling?

When did people begin upcycling?

Try a BIG QUESTION: How does upcycling help the Earth?

SEARCH FOR ANSWERS

Search the library catalog or the Internet.
A librarian, teacher, or parent can help you.

Using Keywords
Find the looking glass.

🔍

Keywords are the most important words in your question.

❓

If you want to know:
- how to upcycle a specific item, type: UPCYCLING [ITEM]
- when upcycling began, type: UPCYCLING HISTORY

LEARN MORE

FIND GOOD SOURCES

Are the sources good?
Some are better than others. An adult can help you. Here are some good, safe sources.

Books
Recyclables by Dana Meachen Rau, 2023.

Trash Craft by Sara Stanford, 2022.

Upcycled Plastic Projects by Heidi E. Thompson and Marcy Morin, 2022.

Internet Sites
Climate Action Superheroes
https://www.un.org/sustainabledevelopment/climate-action-superheroes/
The United Nations (UN) is made up of many countries who have pledged to help fight climate change.

DIY Upcycling Activities
https://www.pbs.org/parents/upcycled-activities
PBS Kids creates educational TV shows for kids. Its mission is to help kids learn and grow.

Every effort has been made to ensure that these websites are appropriate for children. However, because of the nature of the Internet, it is impossible to guarantee that these sites will remain active indefinitely or that their contents will not be altered.

SHARE AND TAKE ACTION

Look for upcycling ideas.
Get an adult's permission to search online. There are many upcycling ideas for everyday items.

Get materials from friends.
You'll need supplies for all your projects. Before heading to the store, ask your friends if they have any materials you can use.

Share your upcycling with others.
A great idea deserves to be shared! You may inspire others to save old items as well.

GLOSSARY

landfill An area where waste is buried under the ground.

material The substance from which an item is made.

plastic A light, strong substance made with oil that can be made into different shapes.

product Something that is made or grown to be sold or used.

reclaim To recover something that was lost.

recycle To turn items back into their raw materials so they can be used to make something new.

sculptor Someone who makes art by carving or molding clay, stone, metal, etc.

thrift store A shop that sells used goods and clothes.

upcycle To change trash into something useful.

INDEX

art 11, 12
clothes 5, 15
Elrington, Ptolemy 13
finding materials 10, 19, 21
furniture 15
landfills 19
problems with upcycling 12
project ideas 10, 11, 16, 19, 20–21
recycling 6, 7, 8–9
selling 14–15
skills 16
thrift stores 11
upcycling clubs 21

About the Author

Amy S. Hansen lives in Maryland. She started writing as a reporter for local newspapers. After a few years she went back to school to study environmental science and then worked as a science reporter. Nowadays, she (mostly) writes for kids because it is much more fun.